Policymakers face certain realities in their efforts to reduce greenhouse gas (GHG) emissions. The effective reduction of GHG emissions implies that the consumption of energy services must reflect the costs associated with their contribution to environmental risks. As well, the regulations and policies implemented to reduce GHGs must be implemented at the lowest cost possible.

These realities have precise implications for all levels of government, including the Quebec government, which is committed to reducing its GHG emissions to the levels set out in the Kyoto Protocol.

Three facts emerge from reviewing the specific situation in Quebec. The first fact is that the possibilities of reducing GHG emissions by replacing one energy source for another are very limited in Quebec. This is mainly due to the reality that the transportation sector depends almost exclusively on oil products – and is a sector with no economically viable fuel replacements, yet. The second fact is that the era of low-cost hydroelectric production is finished. And the third is that the existing hydroelectric facilities provide the most profitable source for exporting clean energy. However, the low hydro rates in Quebec, compared to neighbouring regions, favour heavy usage and limit export capacity.

Based on these facts, two significant measures can be adopted by Quebec in order to meet the objectives of the Kyoto Protocol at the best possible cost. The first measure is a higher carbon tax applied to all sectors. The second is a gradual but significant increase in the price of electricity that would reflect the opportunity cost associated with its increased export value. The joint implementation of these two measures would reduce total domestic energy consumption and allow an increase in clean energy exports. Although these measures would increase the cost of energy in Quebec, access to hydroelectric resources must not impede the adoption of effective policies to participate in world efforts to reduce GHG emissions.

Quebec Policies Concerning Energy and Climate Change

The Quebec government published a May 2006 update of its energy policy entitled, *The Energy to Build Tomorrow's Quebec.* It outlined several priorities, including to:

- Accelerate the hydroelectric development through the implementation of 4500 MW of new projects before 2012 at a cost of $25 billion;
- Develop wind energy capacity to reach 4000 MW in 2015;
- Use energy more efficiently, with reduction objectives ranging from 5 percent to 10 percent, depending on the sources, before 2015;
- Spur energy innovations through support programs.

A month later, the Quebec government revealed a second policy statement dealing specifically with climate change, *Quebec and Climate Change, A Challenge for the Future.* It endorses the average objective adopted for industrialized countries in the Kyoto Protocol; that is a 6 percent reduction of its GHG emissions compared to 1990. That means a decrease of 7.6 percent in CO_2 emissions compared to their level of 84.7 million tons in 2006 (Quebec 2008).

The government identifies two electricity generation technologies with low GHG emissions to carry out this objective; namely, hydroelectricity and wind. A royalty on the carbon emitted by fossil fuels – coal, oil products, natural gas and natural gas fluids – is also currently set at $3/ton of CO_2, and its revenues serve to support initiatives to improve energy efficiency and reduce GHG emissions.

The authors would like to thank anonymous reviewers and Colin Busby for helpful comments on earlier versions of this paper.

Energy in Quebec

In order to better understand the issues in Quebec's new policies concerning energy and GHG emissions, it is important to have a global view. Quebec differentiates itself from the rest of Canada by using more oil and, in particular, more electricity. The relative availability of primary energy in Quebec and in the rest of Canada supports this fact. In 2007, Quebec's reliance on various energy sources compares with that of the rest of Canada as follows: coal, 1.1 percent – Quebec (15.8 percent – Canada); crude oil, 49.5 percent (36.1 percent); natural gas, 12.4 percent (35.7 percent); natural gas fluids, 0.7 percent (5.4 percent) and electricity, 36.2 percent (8.4 percent). In Quebec, electricity is the only local source of primary energy, and all fossil energies are imported. The rest of Canada – mainly the three most western provinces – is a major producer of these fossil energies and a net exporter of them too.

Ninety-six percent of the Quebec production of electricity is hydro. This is an exceptional contribution in global terms. With 212 Tera-Watt-hours (TWh) in 2006, Quebec occupies 4th place as a producer of hydroelectricity, behind China, Brazil and the United States, but ahead of Russia and Norway. Canada, including Quebec, occupies second place. But hydroelectricity represents only 16 percent of the world's electrical production; coal dominates with over 41 percent.

Hydroelectricity emits few GHGs: 60 times less than coal, 40 times less than oil and 20 times less than natural gas (Cliche 2007). The intensive use of hydroelectricity in Quebec ensures that its GHG emissions are lower than in the rest of Canada (Table 1). In 2006, GHG emissions per person in Quebec were 10.7 tons of CO_2, while they were 15 tons in Ontario, 69.5 tons in Alberta and 72.9 tons in Saskatchewan. The emissions intensity by dollar of gross domestic product is also the lowest in Quebec. Therefore, the portion of Canadian emissions originating in Quebec is relatively low; Quebec and Saskatchewan have comparable total emission levels, while Quebec's population is eight times larger.

Table 2 shows the relative distribution of GHG emissions by sector in Quebec and in Ontario. The most significant discrepancies relate, on the one hand, to the share of emissions from the production of electricity, which favours Quebec, and, on the other hand, to the share of emissions from the transportation sector, which favours Ontario. This is the share related to total GHG emissions. Table 1, on the other hand, indicates that Ontario emits twice as much GHGs as Quebec.

Given its status as a public company, Hydro-Quebec charges electricity prices that are among the lowest in the world (Table 3). In the residential sector, the price in Toronto exceeds that of Quebec by 71 percent, while the prices in Boston and New York are 224 percent and 250 percent higher, respectively. The discrepancies are equally significant in the industrial sector. Only British Columbia and Manitoba, which also make great use of hydroelectricity, have lower rates than Quebec.

Quebec's low electricity rates have favoured its usage in all sectors. Over 80 percent of residences are heated with electricity. Half of the demand for electricity comes from the industrial sector, mostly pulp and paper (25 percent) and metal casting and refining (50 percent). With the exception of a plant in British Columbia, all Canadian aluminum smelters, which provide 10 percent of the world capacity, are located in Quebec – the major explanatory factor is the low price of electricity. In 2006, Quebec occupied first place in the world in terms of electrical consumption per inhabitant. It is thanks to this heavy consumption of hydro-electricity that Quebec shows enviable results in total GHG emissions.[1]

While Quebec's electrical network is interconnected with that of its neighbours, annual net exports represent only 5 to 10 percent of available capacity, and they are low compared to the demand of each neighbouring region,

1 See Bataille et al. (2009) for a comparable result concerning GHG emissions by pulp and paper and metal production for all of Canada.

Table 1: GHG Emissions by Province – 2006

	B.C.	AL.	SASK.	MAN.	ONT.	QC	N.B.	N.S.	P.E.I.	NFLD and L.
GHG per person (tons/population)	14.4	69.5	72.9	18.0	15.0	10.7	23.9	21.0	14.9	18.4
Intensity of emissions (kg/G.D.P.)	458	1609	2275	618	423	362	907	828	640	666
Share of Canadian emissions (percent)	8.8	32.9	10.1	3.0	26.7	11.5	2.5	2.8	0.2	1.3

Source: Environment Canada, "Greenhouse Gas Emissions in Canada: Includes trends, 1990–2006," November 2008.

Table 2: Distribution of GHG Emissions by Sector (percent)

	Quebec[a]	Ontario[b]
Transportation	38.7	30.6
Industry	30.7	26.2
Residential, Commercial and Institutional, including Agriculture	21.4	16.9
Waste	7.5	9.2
Electricity	1.6	17.0

[a] In 2005.
[b] In 2004.
Source: Provincial environment ministries.

Table 3: Price of Electricity (Montreal = 100)

	Residential[a]	Industrial[b]
Winnipeg	96	76
Vancouver	99	98
Montreal	100	100
St. John's	155	148
Regina	163	128
Moncton	164	143
Toronto	171	184
Edmonton	173	219
Halifax	176	149
Boston	324	312

[a] for 1000 kWh per month.
[b] for 3060000 kWh per month and a power of 5000 KW Tariffs in effect on April 1, 2008
Source: Hydro-Quebec.

especially at peak times. For example, the interconnection capacity between Quebec and Ontario constitutes less than 5 percent of the peak demand of the latter province, and less than 3 percent of the capacity available in Quebec. In spite of opening markets, the electricity industry is still regional in nature due to constraints imposed by interconnections. These constraints also make it more difficult to export energy, which would further help concentrate efforts on lower-cost options for domestic emissions reduction.

Limited Possibilities for Substitution

Certain realities regarding energy consumption in Quebec cannot be ignored. First, Quebec benefits from exceptional hydroelectric resources and low electricity rates have favoured an intensive use of these resources in all sectors, with the exception of transportation. Second, the use of oil products is widespread in the transportation sector – a sector that still depends almost exclusively on this source of energy – as well as to a lesser extent in the production of heat. Third, Quebec uses little coal and natural gas.

These factors lead to the conclusion that the possibilities of an effective reduction of GHG emissions through the *substitution* of one energy source for another are limited in Quebec. GHG emissions occur mainly through the use of oil products. There are substitutes for heating, but not in the transportation sector where oil products occupy a dominant position. The marginal cost of reduction – that is, the cost of reducing an additional unit – of GHG in the transportation sector, through the substitution of electricity for oil products, is high. From the point of view of effectiveness, it is therefore preferable not to promote this substitution artificially; a well-designed carbon tax would, however, promote a reduction in the consumption of energy devoted to transportation and the replacement of energy sources that issue a great deal of GHG by other sources issuing less.[2]

As mentioned above, Quebec expects to reduce its GHG emissions mainly by accelerating the development of renewable electricity sources, such as hydroelectricity and wind energy. It further expects to reduce the demand for energy by improving energy effectiveness, in addition to encouraging forms of energy that emit less GHGs.[3] The Quebec government has rapidly implemented its new policies. La Romaine hydroelectric project on the Lower North Shore (with a capacity of 1500 MW) has already received the necessary authorizations from the Quebec and federal governments. The work has started, and an agreement has been reached with the native communities affected by this project. Several other projects will allow Quebec to reach its target of 4000 MW of wind energy in 2015. They include: wind farms that are already in operation, the current construction of 1000 MW of wind energy, as well as invitations to tender for an additional 2000 MW, which were made public in the summer of 2008 and which are already approved.[4]

Evaluation of the Quebec Policy on Climate Change

On the international stage, Quebec, together with Ontario and New Brunswick, is an observing member of the Regional Greenhouse Gas Initiative (RGGI), which includes the states of New England as well as New York, New Jersey, Delaware and Maryland. Similar to British Columbia, Manitoba and Ontario, it is also a partner of the Western Climate Initiative (WCI), which was launched under the leadership of California, and mainly involves states from the US West Coast. These groups share the same objective of reducing GHG emissions through the implementation of a tradable permit system. In the summer of 2008, Quebec and Ontario decided to launch a tradable permit system in order to prepare the ground for a national system. The Montreal Climate Exchange (MCX) launched its operations in May 2008,

2 A GHG tax would for instance, through shifting relative prices, promote the development and use of rechargeable hybrid vehicles.

3 The Quebec strategy also relies on the substitution towards non-emitting energy forms, such as biomass and biofuels.

4 The Energy Efficiency Agency has already submitted its global energy efficiency plan to the Energy Department and this plan is funded by a share of the energy distributors, including electricity.

offering term contracts on carbon dioxide (CO_2e) equivalent units.[5]

To evaluate the quality of Quebec's efforts to reduce GHG emissions, we should first acknowledge that decreasing GHG emissions is desirable for all humanity and for future generations. Clearly a public good in economic terms, a reduction in the GHG emitted by a country affects the global level of emissions.[6] It is obvious that the impact of the actions adopted by Quebec in this regard can only be marginal – its GHG emissions constitute less than half a percent globally. However, Quebec can still contribute a reasonable share to the reduction of GHGs at the global level. The main criterion to evaluate its efforts, then, is their effectiveness in achieving this reasonable share at the lowest cost possible.

Canada signed the Kyoto Protocol in 1997 and ratified it in 2002. It has committed itself before the international community to reduce its GHG emissions by an average of 6 percent, compared to the level of emissions in 1990, between 2008 and 2012. After discussions, policy statements and plans, very little progress has been achieved: the current level of Canadian emissions exceeds the planned objective by more than 30 percent. In 2007, the Canadian government unveiled the *Regulatory Framework for Air Emissions*, a program focused on the intensity of emissions by large industrial emitters. The target: to reduce their emissions by 20 percent by 2020, compared to 2006 levels. This is a very modest goal compared to the commitment assumed under the Kyoto Protocol (Jaccard 2007). This plan relies essentially on regulating large emitting industries and, with the exception of a complex package of cap-and-trade measures, makes little use of market mechanisms. It sets targets for average carbon use by production unit, but not total emissions. It further includes a set of exemptions and special treatments that renders the efforts to reduce greenhouse gas emissions costly, ineffective and arbitrarily distributed. Contrary to a carbon tax, the plan does not implement mechanisms that lead to equalization, through companies and industries, of the marginal costs of the reduction of GHG, as would be called for on efficiency grounds. As it stands, the Regulatory Framework makes it difficult to achieve inter-industrial, interprovincial and inter-temporal coordination of GHG emissions reduction.

The new American president is more favourable to the reduction of GHGs than his predecessor, who refused to ratify the Kyoto Protocol. Until now, Canada has mainly had this file on hold, waiting to see the policy adopted by other countries, in particular, the United States. However, the Canadian government has recently announced its desire to implement a cap-and-trade system for North America.

Meanwhile, the fact that the environment constitutes an area of shared jurisdiction for Ottawa and the provinces has allowed the latter to act according to their own interests with regard to policies for the reduction of GHG. WCI involvement by provinces that are rich in hydroelectric resources, such as Quebec, British Columbia, Manitoba and Ontario[7] – and the noticeable absence of Alberta – are not completely accidental. These provinces would benefit from an increase in the cost of electricity produced from fossil energies, because this would lead to a higher value for their exports. Quebec is apparently getting ready for expected increases in electricity prices in its neighbouring jurisdictions by increasing the capacity of some interconnections. Work is already in progress to increase the trade capacity with Ontario by 1250 MW, and there have been discussions with New England in this regard.

This leads us to wonder about the source of electricity to feed the neighbouring networks. It is true that Quebec is developing, or plans to develop, new resources. La Romaine hydroelectric project has an estimated cost of 10¢/kWh. The latest invitation to buy 2, 000 MW of wind

5 See press release at http://www.m-x.ca/f_comm_press_fr/27-08_fr.pdf.

6 This does not mean that a decrease in the GHG emissions would necessarily have beneficial effects that would be equally distributed.

7 In 2007, Ontario produced 50.8 percent of its electricity from nuclear energy, and 21.5 percent from hydroelectric resources, which is less than these other provinces. These two electricity-producing processes emit little GHG jointly.

energy was settled for an average price of 10.3¢/kWh. By comparison he average price received by Hydro-Quebec for interruptible electricity exported to the United States was 8.3¢/kWh in 2008.[8] There will therefore have to be a significant increase in the price of electricity in the American Northeast to make Quebec's new investments profitable. Because Quebec's hydroelectric sites have been developed in sequence with low-cost production coming on stream first, the low-cost hydroelectricity era appears over in Quebec.

In summary, Quebec, which has an exceptional supply of hydroelectric resources, is about to develop resources that are just marginally profitable in order to export clean electricity. Alternatively, it could increase the domestic price to better reflect: 1) the cost of new supplies; 2) the market price in neighbouring regions; and, 3) the importance of reducing GHG in the most effective way possible. Just because Quebec has major hydroelectric resources, it should not forget the basic rules concerning the efficient use thereof. Having clean energy at a low cost does not mean that you are economically justified in wasting it. The possibilities of producing clean energy elsewhere are limited and costly, whether it is wind, solar or hydroelectricity. Any overconsumption of energy reduces its export capacity and increases the costly and harmful production of less clean energy elsewhere.

Conclusion and Recommendations

Quebec states that it supports the initiatives aimed at reducing GHG emissions and plans to benefit from the resulting increases in electricity prices experienced in neighbouring markets. Nevertheless, internally, regulations and subsidies are the tools currently preferred to participate in this collective effort. Quebec, then, has a strangely asymmetric position with regard to controlling GHG emissions. On one hand, it favours the use of market mechanisms outside its borders; on the other hand, it has resorted to regulation and subsidies to influence domestic energy consumption. In a recent speech, Premier Jean Charest expressed the desire that "sustainable development and the fight against climate changes [be] synonyms with prosperity." Quebec will achieve that goal more easily if it submits the development of its energy wealth to a more coherent and wider use of the market mechanisms that will allow it to reduce GHG emissions at the best possible cost for all the Quebec society.

Two major uses of such mechanisms seem particularly relevant: a gradual and considerable increase in the price of electricity that will better reflect its actual cost; and a higher carbon tax to promote the consumption of clean energy.

As we have seen, the electricity produced from existing facilities will be the most profitable to export for Quebec. The prospects of increased exports are, however, limited by domestic electricity consumption, which is supported by particularly low electricity prices. These low prices are the result of rate regulations that are based on average historical costs, which are clearly lower than the costs of new projects. Therefore, a gradual and considerable increase in the price of electricity should be implemented in order to promote a reduction in consumption and support increased exports of clean energy.[9] This would obviously result in an increase in the cost of electricity for everyone, but the increased revenue for the state would allow the implementation of redistribution policies to lessen the impact on the less wealthy (Boyer 2005). Such an increase would also tend to increase energy efficiency, a goal the government is currently trying to achieve through regulations and subsidies. A change in approach would allow Quebec to export more electricity to take part in global GHG reduction efforts.

As mentioned above, a small royalty on carbon emissions associated with fossil energies was introduced in 2007, at a rate of $3/ton of CO_2.

8 See www/neb/gc.ca.

9 Bernard and Genest-Laplante (1995) show price elasticity estimates in the sectional demand of electricity as higher than one.

This carbon tax should be raised to reflect the actual and increasing cost of GHGs. On the low end of suggested GHG prices, the current price on the RGGI market for a permit to emit one ton of CO_2 in December 2009 is $4.15. On the upper end of GHG prices, the Stern report on the effects of climate change recommends a carbon tax of $90 per ton (Stern 2006). Because a carbon tax of about $3 per ton is equivalent to a gasoline tax of 1 cent per litre, the cost range of GHGs suggests that a carbon tax from $15 to $90 per ton requires a tax increase from 5 cents to 30 cents per litre of gas.

Therefore, an appropriate measure for Quebec is to announce a gradual increase in the tax on CO_2 emissions, similar to that introduced in British Columbia. British Columbia introduced a $10 tax per ton of CO_2 in 2008, which will be increased by $5 annually to reach $30 in 2012.[10] Quebec's tax, which is currently at $3/ton of CO_2, should be increased annually by $3, in order to reach a target price of $15 in 2013 and $30 in 2018. A more aggressive approach would be to increase the tax to $30 per ton in 2013; that is, a year later than British Columbia reaches the same $30 per ton target. The energy substitutions and reductions carried out by consumers would result in lower and cleaner energy consumption, in accordance with the objectives of reducing GHG.

Gradually increasing the price of electricity and the carbon tax is the best approach to reducing GHG emissions in Quebec. The political resistance to an increase in the price of electricity could be reduced with a gradual increase in tariffs and through protection mechanisms for low-income consumers. Better electricity prices and a revamped carbon tax would have the additional advantage of preparing the Quebec economy for an active and influential participation in a North American program for the reduction of GHGs – a program that might be soon implemented by Ottawa and Washington. These measures would favour a smoother and better-anticipated reduction of carbon consumption in Quebec, and a more beneficial involvement for Quebec in any federal or North American cap-and-trade system.

10 The bill on the control of GHG emissions adopted by the Chamber of Representatives of the United States in July 2009, includes a floor price of $10/ton of CO_2 in 2012 and this floor price will increase at an annual rate of 5 percent. A ceiling price will also be implemented at the initial level of $35/ton and it will increase at a rate of 4 percent per year.

References

Bataille, Chris, Benjamin Dachis, and Nic Rivers. 2009. *"Pricing Greenhouse Gas Emissions: The Impact on Canada's Competitiveness."* C.D. Howe Institute Commentary 280. February.

Bernard, Jean Thomas, and Eric Genest-Laplante. 1995. *Les élasticités-prix et revenu des demandes sectorielles d'électricité au Québec: revue et analyse.* Final research report submitted to Hydro-Quebec.

Boyer, Marcel. 2005. "Augmentons le prix de l'éctricité au Québec – pour le bien de tous" C.D. Howe Institute e-brief. C.D. Howe Institute: Toronto. March.

Cliche, Yvan. 2007. "L'Hydroélectricité: une solution aux changements climatique" Hydro-Quebec: CHOC Magazine. October [http://www.aieq.net/_site/documents/applications/pdf/ChocOct07-Solution.pdf] (August 12, 2008).

Jaccard, Mark. 2007. "Designing Canada's Low-Carb Diet: Options for Effective Climate Policy." C.D. Howe Institute Benefactors Lecture. November.

Stern, Sir Nicolas. 2006. *Stern Review on the Economics of Climate Change.* Final Report to HM Treasury. London.

Quebec. 2008. "Mise à jour du Plan d'action du Québec : Le Québec et les changements climatique" Government of Quebec: Ministère du Développement Durable, de l'environnement et des Parcs. June. [http://www.mddep.gouv.qc.ca/changements/plan_action/2006-2012_fr.pdf] (Date of Access: August 12, 2008).

C.D. Howe Institute Backgrounder© is a periodic analysis of, and commentary on, current public policy issues. James Fleming edited the manuscript; Heather Vilistus prepared it for publication. As with all Institute publications, the views expressed here are those of the authors and do not necessarily reflect the opinions of the Institute's members or Board of Directors. Quotation with appropriate credit is permissible.

To order this publication please contact: Renouf Publishing Company Limited, 5369 Canotek Road, Ottawa, Ontario K1J 9J3; or the C.D. Howe Institute, 67 Yonge St., Suite 300, Toronto, Ontario M5E 1J8. The full text of this publication is also available on the Institute's website at www.cdhowe.org.